Once there were two friends who were very fond of sailing. Every day, right after lunch, they went for a sail in their little boat. It was a nice little boat with seats, a basket to keep cookies in, a flowerpot for emptying out water, and a beautiful umbrella to shade them from the sun.

1

One afternoon the two friends came strolling down to the boat, carrying their cookie basket. Suddenly they stopped.

"The umbrella!" cried Mr. Woolsey. "It's gone!"

"Someone has stolen our umbrella!" said Mr. Tootle.

"Oh, no!" said Mr. Woolsey. "Now we can't go sailing."

So they picked up their cookie basket and went home.

Sometime later while they were having tea, Mr. Tootle cried, "We must go back! We forgot to look for CLUES!"

"You are so right," said Mr. Woolsey.

And the two friends left their tea and hurried back down to the boat. But—

There was the umbrella, right where it belonged!

"The thief has brought our umbrella back," said Mr. Tootle. "He must have felt guilty."

"You are so right," said Mr. Woolsey. "But it's too late to go sailing now. We'll have to wait until tomorrow."

The next day, right after lunch, the two friends went hurrying down to the boat. But—

"The umbrella!" cried Mr. Woolsey. "It's gone again! What kind of a thief can this be, to play games with two such upstanding birds?"

"Stop the talk," said Mr. Tootle, "and look for clues."

They began to look in the tall grass. Suddenly—

"CLUES!" cried Mr. Tootle. "A red one, right over your head!"

Mr. Woolsey looked up, and there, stuck in the high grass above his head, was a little round ball hanging like a bright red berry. It was one of the dingle-dangles that he had sewn around the edge of the umbrella to make it look cheerful.

Mr. Tootle pulled the ball off the grass and put it in the cookie basket. "We're on the right track," he said. "Let's see what else we can find."

They searched for some time. But they missed the cool shade of their umbrella. "We should go home now," said Mr. Tootle, "and come back later. The heat of the day is a fine time for sailing, but it is no time to go looking for clues."

"You are so right," said Mr. Woolsey. And off they went for home.

Later that day, they came back down to the boat.

"The umbrella!" cried Mr. Woolsey. "It's back again. This *is* upsetting!"

"It's upsetting because it's so strange," said Mr. Tootle. "But—" He scratched his head. "The fact is, our umbrella is always missing when we want to use it, which is at noontime in the heat of the day. And it is always back in place when we don't want to use it, which is all other times of the day. So—whoever takes our umbrella, only takes it—"

"—when WE want it!"
"Stop interrupting. That's not what I was going to say. He takes it when the sun is hottest! We'll make a lunch and come down early . . ."

Mr. Woolsey wasn't listening. He was thinking his own thoughts.

"I hope the thief won't keep dragging that umbrella through the tall grass," said Mr. Woolsey. "I can just see a picture of that grass in my mind's eye with little red balls everywhere—like a field full of red lollypops!"

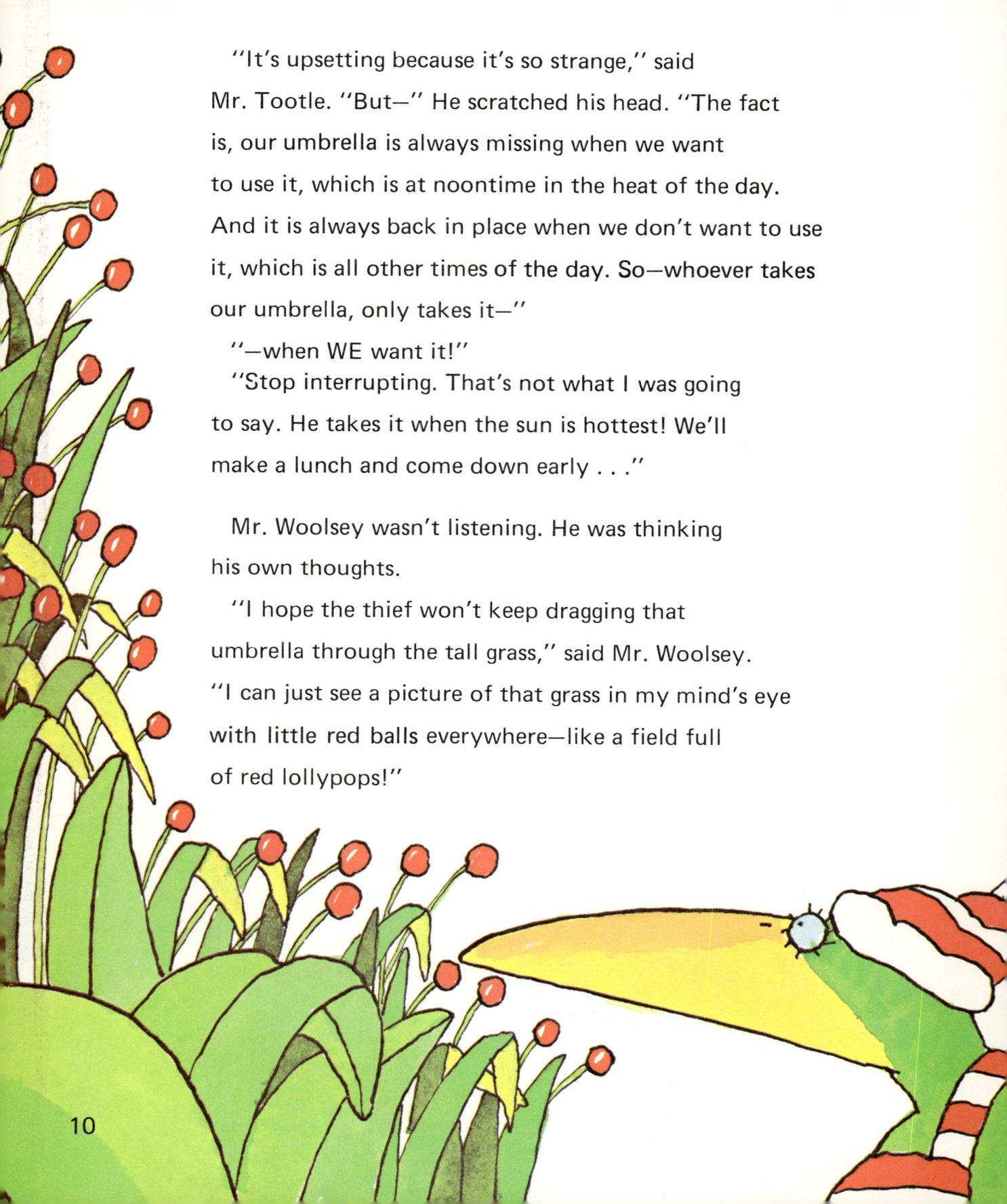

"Stop talking nonsense," said Mr. Tootle. "I was having an idea, and you never even noticed. I said we'll come down early and hide. Then we'll see who is taking our umbrella. We might even follow him and see what he does with it!"

"You are so right," said Mr. Woolsey. "It's a good thing we are two such smart birds!"

The next morning the two friends hid in the grass. They waited and waited. Dragonflies flew over the water. The sun climbed higher in the sky. Then, the tall grass parted. A head popped out and looked around. And Archibald Turtle stepped onto the path. He went straight to the boat, lifted the umbrella, and started off with it.

The two friends followed.

They went along the edge of the water. At last the grass gave way, and they came to a clearing. The two birds watched.

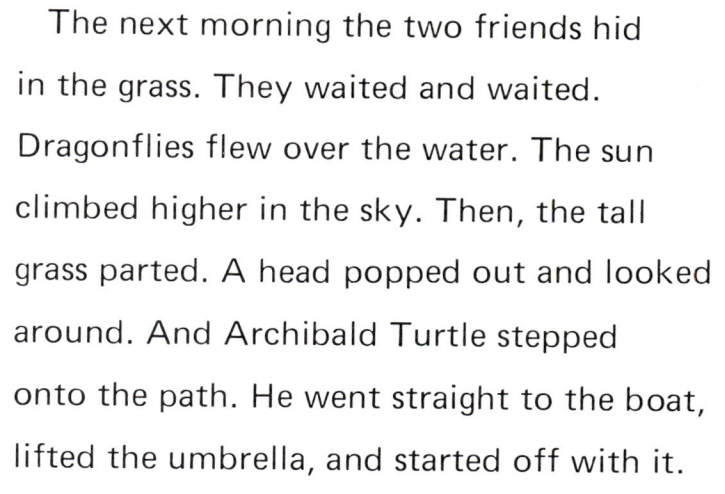

In the clearing was a garden. It was beautifully kept and edged with shells. There was nothing in it, except for one small plant. Archie carefully placed the umbrella to make a circle of shade over the plant.

"Well, I never!" whispered Mr. Woolsey. "He can't be such a bad thief, since he takes it for a good cause."

"A thief is a thief!" said Mr. Tootle.

"But he always brings it back."

"Just the same," said Mr. Tootle, "it's our umbrella!"

And he stepped out into the clearing.

"See here, Archie. That is our umbrella, and we need it for our boat." Mr. Tootle went to pick it up.

Archie grabbed the umbrella and held it in place.

"No, no," he cried. "My marigold is sick! Can't you see? It needs to be in the shade."

"But it's our umbrella," said Mr. Tootle. "It belongs in our boat."

"I have it!" cried Archie. "Bring your boat over here. Then you can shade your boat and my marigold at the same time. You can use my wagon."

"NO!" cried Mr. Tootle.

"Aren't we being a bit selfish?" asked Mr. Woolsey. "It couldn't hurt anything to bring the boat over here."

"Come to think of it," said Archie, "I'd be doing you a favor to let you use my wagon. After all, it is *my* wagon."

"Well, I suppose—" said Mr. Tootle, shaking his head. "But it's against my better judgment."

"You are so right," said Mr. Woolsey.

The three of them took the wagon and went back to the boat. They pushed the wagon into the water, floated the boat on top of it, and pulled. The boat came slowly out of the water on top of the wagon.

They had nearly reached Archie's garden when Mr. Tootle stopped.

"It won't work," he said.

"Why not?" asked Mr. Woolsey.

"Well, it's true we don't ever go anywhere when we go sailing, but still, you can't go sailing in a garden. You *have* to have water." He frowned at the others. "If you'd thought things through, you would have thought of that."

Nobody said anything. Then—

"I know," said Mr. Tootle, looking pleased. "There's nothing so wrong with what we are doing; it's only that we are doing it backwards. It's the marigold that should go into the boat!"

"Oh, you are so right," cried Mr. Woolsey. "It certainly helps to think things through. We even have a flowerpot right here in the boat. It never was much good for emptying out water."

They left the boat and went back to the garden. Archie dug up the marigold and got his watering can. Mr. Tootle picked the umbrella, and Mr. Woolsey the cookie basket. They all set off for the boat. By the time they had it back in the water, the sun was getting low in the sky.

"It's a little late," said Mr. Tootle, "but if we hurry, we'll just have time for a short sail."

"You are so right," said Mr. Woolsey.
There was just enough room in the boat for Mr. Tootle, Mr. Woolsey, Archie, the sick marigold, the watering can, *and* the cookie basket. They stayed until the sun grew low in the sky. Then they went home.

Now there are three friends who are
very fond of sailing. Every day, when the sun
is high in the sky, they go down
to their little boat. It is a nice little boat
with seats, a cookie basket, a marigold,
and a beautiful umbrella to shade them
from the sun. Mr. Woolsey sewed
the dingle-dangle back on the umbrella,
and the three friends are very happy.